TOM BRADLEY, mayor of Los Angeles and moderator of the Round Table: I am delighted to welcome to Los Angeles AEI's National Energy Project and this important discussion of the energy problem. In my view, it is appropriate that we should come to Los Angeles to talk about "Offshore Oil, Costs and Benefits."

We have a distinguished panel to discuss this matter: First, Mr. H. J. Haynes, chairman of the board and chief executive officer of the Standard Oil Company of California; second, Captain Jacques-Yves Cousteau, internationally known explorer, philosopher, and poet, and a distinguished television performer and producer who opened up for millions the world under the sea. Also on our panel are the governor of New Jersey, the Honorable Brendan Byrne, and Assistant Secretary of the Interior Royston Hughes.

Governor Byrne, would you begin the discussion this morning?

BRENDAN BYRNE, governor of New Jersey: The subjects under discussion here are the costs and benefits associated with offshore oil and gas development. Usually the governors of the coastal states do not see eye to eye on any subject, but on this subject, I think, there is close to unanimous agreement within our group that we ought to proceed with development only after gaining full knowledge as to its implications. We want to know what will be going on if the development of offshore oil occurs.

Let me state how I and the other coastal governors

would go about the development of a resource under the ocean near our coastlines, especially the coastlines of states sensitive to development. If we did not know what our federal government was doing, we would try to get as much information as possible from it. We would ask the oil companies to pool their resources to gather what information they could and provide it to us. Then, using all that information, with whatever we could get from our federal government, we would proceed very cautiously, but in an orderly step-by-step manner, using a rational plan to develop those offshore oil and gas resources. We would want to know what the effects would be on the ocean, what the effects would be onshore, and what would be the best way of minimizing or eliminating those effects.

In contrast, the worst possible thing to do would be to divide the offshore region into a series of huge tracts and put those tracts out for bidding, and then to say that the highest bidder would get a shot at both exploring and developing those tracts without any additional scrutiny. In other words, once the bid was out, good luck. I think the federal government is using a procedure a lot closer to this than to the rational plan, and many of us—many of the governors of the coastal states—have raised our voices against that.

I think we have made some progress. I think Mr. Hughes and Mr. Morton are beginning to see the dangers in putting huge areas out for bidding under the current scheme. I would emphasize that the governors are willing to see offshore oil and gas developed rationally, in a way that will protect the citizens of our states and of the United States. We want to see an optimum result for the best interests of all our people. We are not obstructionists, only realists.

MAYOR BRADLEY: Thank you very much, Governor Byrne. Mr. Haynes, would you take up on this issue, perhaps from a different point of view?

H. J. HAYNES, chairman of the board, Standard Oil Company of California: Discussions of the subject of offshore drilling have, to say the least, become somewhat

OFFSHORE OIL: COSTS AND BENEFITS

Tom Bradley, *Moderator*

Brendan Byrne
Jacques-Yves Cousteau
H. J. Haynes
Royston Hughes

An AEI Round Table held on March 21, 1975
and sponsored by
the American Enterprise Institute for Public Policy Research
Washington, D.C.

THIS PAMPHLET CONTAINS THE PROCEEDINGS OF
ONE OF A SERIES OF AEI ROUND TABLE DISCUSSIONS.
THE ROUND TABLE OFFERS A MEDIUM FOR
INFORMAL EXCHANGES OF IDEAS ON CURRENT POLICY PROBLEMS
OF NATIONAL AND INTERNATIONAL IMPORT.
AS PART OF AEI'S PROGRAM OF PROVIDING OPPORTUNITIES
FOR THE PRESENTATION OF COMPETING VIEWS,
IT SERVES TO ENHANCE THE PROSPECT
THAT DECISIONS WITHIN OUR DEMOCRACY WILL BE BASED
ON A MORE INFORMED PUBLIC OPINION.
AEI ROUND TABLES ARE ALSO AVAILABLE ON
AUDIO AND COLOR-VIDEO CASSETTES.

ISBN 0-8447-2076-3
LIBRARY OF CONGRESS CATALOG CARD NO. L.C. 76-3350

PRINTED IN UNITED STATES OF AMERICA

emotional and more than somewhat confusing. The fact is that this nation is consuming more oil than it is producing: in 1950 we were importing practically no oil from foreign sources, while this year we will probably be dependent on foreign sources for 40 percent of our oil requirements, and unless something is done, by 1977 or 1978 we will be importing perhaps 50 percent of our oil requirements.

I was reading an article in one of the major newspapers the other day, and two points mentioned in the article struck me as quite pertinent. One is that this nation is spending as much money today on imported oil as it was spending on the Vietnam War at its peak. The second was that if we were to undergo another embargo such as the one that occurred in the winter of 1973-74, it could mean the loss of 2 million jobs in this country and could affect our gross national product by some $50 to $80 billion. I think that everyone on this panel will agree that, given those facts, we should get on with some kind of a program to minimize our dependence on foreign sources of energy. I think everyone will agree to that.

Now, there is no question that our country has an abundance of alternate energy resources. Certainly we have an abundance of coal; we have the potential to expand our nuclear generating capacity; we have the potential to gasify coal and to liquefy coal; we have tremendous shale oil resources. Certainly the development of all of these alternate resources should be pursued, and the oil industry is willing to play its part in the development of these alternate resources, but the industry—like all of us—needs an agreed-upon national energy program that will provide a stable economic and political climate within which to get this job done.

Having said that, I think I should point out that in my judgment all these alternate energy resources will not play a substantial role in solving our energy problem, certainly not in the next decade and perhaps not for as long as fifteen years. I do not see any alternatives in the short term, in the next decade, to exploring and developing our domestic oil and gas resources. The experts in this field have made studies—the U.S. Geological Survey, the National Petroleum Council (NPC), the National Academy of Sciences (NAS)—and while the various experts differ in their estimates of our potential resources, they all believe that the greatest potential for additional oil and gas in this country lies in the outer continental shelf (OCS) and in Alaska.

I am, of course, part of the energy industry, and I think I have a responsibility to the American public to advocate what in my judgment is really the only feasible short-term solution to this problem. For me to take any other position would, I think, be totally irresponsible, and for that reason, I urge the federal government and the states to get together and work out their differences—and while they are working out their differences, let us get on with exploring for oil offshore, because offshore exploration really is the only short-term solution.

Many of the people in this country think of energy as the means of getting from where they might be to where they want to go, or perhaps as turning a light on at home or something of that sort. But I would like to emphasize the fact that energy is the underpinning of our economic system—which means that energy can be equated to jobs. The offshore drilling decision is basically a political decision, but I hope that the decision makers, when they deliberate and establish priorities, will recognize that we are not just talking about inconveniences—about getting places or turning out lights—when we talk about a shortage of energy in this country. We are talking about jobs and an adverse effect on our economy.

JACQUES-YVES COUSTEAU, The Cousteau Society: First, I am a little embarrassed to be here on this panel as a foreign citizen. My only reason for accepting the invitation is that I am paying my taxes in the United States; otherwise, I would have turned this invitation down. Also, I do spend most of my time here when I am not at sea.

MAYOR BRADLEY: You have earned your right to speak.

CAPTAIN COUSTEAU: I am very disturbed that throughout this conference, as we have discussed the costs and benefits of offshore drilling, our starting point has always been the assumption that offshore oil is definitely needed. This need has never been substantiated here by studies or figures; it has merely been assumed as an axiom. Now I am not saying that offshore drilling is not needed. I am just saying that during this conference the need has not been demonstrated.

A second thing. Looking at this problem from a double point of view, both as an ocean explorer and leader of scientists—though I am not a scientist myself—and as a man who has time to think when up on the bridge, I have reflected that everyone is talking about short-term independence, and it occurs to me that desire for short-term energy independence for this country may be more dangerous than temporary dependence and long-term independence. Let me explain what I mean.

Of course, we know and the Europeans know very well what it costs to buy energy outside. Here you buy perhaps 40 or 50 percent of your oil from abroad. In France we buy 90 percent from the Arab oil countries, and the present cost is a terrible drain on our economy.

But we are in a time when international affairs are relatively at ease, even if the headlines in some papers dramatize things. It is said that things have rarely been as easy as they are now for the Western world. Certainly this is unlikely to last forever. In ten or fifteen or twenty years we may face some critical strategic situations for which easy oil energy will be required. If out of expediency we have quickly burned our last offshore reserves in time of peace and ease, we will be without any kind of resources in hard times, and we will be strangled. This possibility has formed my general approach to this problem. I know that offshore drilling is technically feasible, and that the protection of the environment can be dealt with in ways that we may talk about later on. I know that, as a temporary bridge to the future, offshore oil may be necessary. Nevertheless, it seems to me that right now we have to develop other sources of energy, especially of the renewable and nonpolluting kind, though it is likely to take enormous funding.

ROYSTON HUGHES, assistant secretary of the interior for program development and budget: It is a pleasure to have the opportunity to explain the government's position on offshore oil drilling. Our topic today is costs and benefits, and I might mention that in 1970 this country spent $3 billion for offshore oil, while last year, 1974, we spent $24 billion, which represents a significant cost to the American public. We believe that among the total array of energy sources available to us for the near term, offshore oil, as Mr. Haynes said, provides the one real additional source of domestic supply.

In this country we use about 17 million barrels of oil per day. We produce about 11 million barrels of oil a day, and our production is declining—last year it declined 8 percent, and perhaps next year it will drop by 6 percent. Certainly, over the short term, over the next ten years, there will be a significant decline in the production of domestic American petroleum. That is one of the reasons we believe that it is incumbent upon the government to ensure that we conduct an orderly program to determine where our additional resources are. Of course we are concerned about the prospect of using up our fossil fuel resources now and not having any fossil fuels to serve us later on, but we really do not believe we are planning to use up our offshore oil.

Our program right now is an orderly series of lease sales over the next four years that will permit exploration of the remaining unexplored areas of the outer continental shelf so we can find where the oil is and where it is not. We recognize that both strategic and domestic production conditions may change so as to force us to revise a given schedule. We believe the government's program has the flexibility necessary for dealing with changes in our energy situation. Even so, because of the long lead times involved in offshore drilling, we believe it is incumbent on the government and certainly on the American public to support a program that will provide us with those resources when we need them.

The government recognizes that, as the program expands from its traditional area of operation in the Gulf of Mexico, we will have to explain it carefully in the new states involved. We recognize that the program will have certain effects on the coastal areas and we are trying to come to grips with the problems that the states may face from offshore development. We have consequently made a series of changes in the program that will meet many of the objections made in the past. I think the most important change—one that I hope we can discuss in a little more detail here—is a commitment on the government's part to work more closely than before with the states. We intend to include state officials in the process by which the ultimate decision is made to lease or not to lease.

We want to continue working on our system to see whether there are additional changes that could make it better. There is a great national debate now on Capitol Hill as to what will ultimately be the national energy policy. As that debate goes on, we must continue to work on the outer continental shelf program. We do not see any other source of petroleum in the short term. Of course, coal gasification, coal liquefaction, solar energy, and geothermal energy are all potential future energy resources; but when we talk about these, we are talking about their becoming available beyond 1985 and 1990, and we need immediate relief from our short-term energy supply problem.

MAYOR BRADLEY: It seems to me that Governor Byrne is saying, "Not so fast, hold up, let's have a little more orderly process." Mr. Hughes is saying, "We think this is a reasonable process, the phasing of this program is well timed, and we want to involve the states." I did not hear him say "involve the cities," but I know he intended to. Captain Cousteau is saying, first, that we have seen no demonstrable evidence in this conference that we ought to be drilling on the outer continental shelf, and, second, that we ought to develop some alternate sources of energy right now, despite the costs. Mr. Haynes is saying that there is only a limited amount of oil available in this country and in the world, and we had better get on with our work because it will take ten or fifteen years even to bring it onshore for refining.

MR. HAYNES: May I correct a misunderstanding? I certainly did not intend to suggest there was a shortage of oil in the world. Indeed, there is a fully adequate supply of oil in the world. Our problem is with the supply of oil right here in the United States, and my view is that we should minimize our dependency on these foreign sources for the many reasons which Mr. Hughes mentioned. There is a considerable amount of oil in the world, but here in the United States we are consuming a lot more than we are producing. I think in the interests of our economy, perhaps also in the interests of our national security, it behooves us to limit our dependence on foreign sources.

Of course, there is no question that oil is in finite supply, but I want it to be clear that in the foreseeable future there is an abundance of oil in the free world. We are all aware of the fact that there are many millions of barrels a day of surplus producing capacity at the present time in the Middle East and among non-Middle Eastern OPEC countries. The supply is finite, but at least there is still quite a lot of oil for the future. Unfortunately for us as Americans, two-thirds of the oil in the free world happens to be located in the Middle East, and a larger percentage in the OPEC countries.

CAPTAIN COUSTEAU: What bothers me is that we seem to be saying that there is a lot of oil elsewhere in the world and we should burn ours quickly in the United States. Do I understand what is being said?

MR. HUGHES: I do not think it can be said that our position is to burn it quickly, since the lead time involved is substantial and the process of exploration I referred to earlier is an orderly one. From the opening of the proceedings to sale of rights takes up to two years. It takes one to three years from the time of the sale of rights to exploration, and three to eight years beyond that before development might occur, should oil be found. I would emphasize that phrase—"should oil be found"—because I believe we all keep it in mind that the oil business is a speculative one.

In any case, the operation of this system now takes years rather than months or days. If we delay now for several years in order to determine whether or not the system could be a little better, ten years from now we might find that we had started five years too late, and we might be faced with a massive change in the American lifestyle.

GOVERNOR BYRNE: The governors have been portrayed as the heavies in this debate. Perhaps we ought to clear that up a little bit with Mr. Hughes and Mr. Haynes. If the governors today suddenly withdrew every objection they had to offshore drilling, where would we be? Where would we go tomorrow?

MR. HUGHES: I think it is our intention to proceed with the orderly process that we outlined. We do not believe the governors are the heavies: after all, there are other groups besides state officials concerned about the conduct of the program. But we do think it is in the national interest to proceed with an orderly plan to develop oil resources and have them available for the American people, and we intend to proceed within the context of existing laws.

GOVERNOR BYRNE: Are we governors lousing up an orderly plan that you had and have, Mr. Hughes?

MR. HUGHES: No, I think that the governors have sharpened the debate, and that is what our system of government is all about.

GOVERNOR BYRNE: Do you have a plan that would let you go ahead tomorrow with accelerated exploration to determine where the energy resources are, and still allow an orderly decision-making process?

MR. HUGHES: It depends on what area you are talking about. If your reference is to the Atlantic, I can tell you that there we intend to resume our scheduled initial call for nominations.

GOVERNOR BYRNE: Your environmental impact statements will let you do that now?

MR. HUGHES: Not yet; the environmental impact statements are part of the process.

GOVERNOR BYRNE: I see.

MR. HUGHES: The decision to lease will not be made until all these things that you refer to are accomplished.

GOVERNOR BYRNE: The Interior Department filed an environmental impact statement in October which attempted to justify the entire speeded-up program.

MR. HUGHES: That environmental impact statement covers the entire program. For each given sale area, the department does a site-specific environmental impact statement, taking into account local environmental, social, and economic factors.

GOVERNOR BYRNE: You do not really expect anyone to believe that the environmental impact statement the department filed would justify offshore drilling today, do you?

MR. HUGHES: We think the statement justifies offshore drilling.

GOVERNOR BYRNE: That statement would justify offshore drilling?

MR. HUGHES: In an overview, covering the seventeen

areas in the outer continental shelf, ranging from Alaska
to the entire east-west coast and the Gulf of Mexico, the
statement justifies offshore drilling.

GOVERNOR BYRNE: You had hearings on that environmental impact statement, did you not?

MR. HUGHES: That is true.

GOVERNOR BYRNE: Some were in Trenton, some here,
others in Alaska and elsewhere. I did not hear of anyone
in that state characterizing that environmental impact
statement as being adequate to justify drilling offshore
today.

MR. HUGHES: I think there were some supporters. I
know it can be said that the only supporters were in the
oil industry, but we believe that the offshore drilling program has a number of supporters greater than the number
of persons who voiced their opposition to the program
at these public hearings.

GOVERNOR BYRNE: I do not care about counting noses.
I do care about whether the environmental impact statement the department has come up with so far provides
an adequate basis on which to proceed. That is the
question.

MR. HUGHES: If your question is whether we will make
some additional changes in the statement, the answer is
that we will. Nevertheless, we think that by and large
the statement is an adequate document.

GOVERNOR BYRNE: You mean that you could have
gotten away with that statement.

MR. HUGHES: We do not like to use that kind of lan-
guage. We think the statement was prepared in good
faith; and that it does in fact meet the requirements of

the National Environmental Protection Act (NEPA) and the general concerns outlined in that act.

MAYOR BRADLEY: Governor Byrne, I think your reputation as a prosecutor is well earned. Let us see if we can get to our other panelists now. A common question is whether we need to proceed as quickly as we are on offshore drilling. Some observers claim that, even if we were to proceed today, we could not extract the oil, we could not bring it in, we could not refine it, in less than eight or ten years or perhaps longer.

MR. HAYNES: This question is almost always put into the form, Should we move so rapidly? I think we ought to ask ourselves instead, Should we continue to delay this process? The oil industry has been asking for accelerated leasing offshore for years. We respect the efforts of the various interested parties, the environmentalists and the states, and we have advocated for some time—we have gone on record to this effect—that there should be some mechanism whereby the coastal states could share in whatever revenue or economic benefit the federal government derives from offshore leasing. But I am concerned when the question is raised only in the form, Should we proceed so quickly? when in fact we have been delayed almost to the point of irresponsibility. I have already said several times that there is a lot of oil in the world, and Captain Cousteau has said that since the world is relatively stable politically right now, there is no reason to think we could not continue to bring this oil in. But I would remind you of a point I made in my opening statement— we are still open to a political embargo on oil.

If the embargo is enforced again, it will not merely be an inconvenience, not merely mean long lines at service stations—though I do not like those any better than anyone else, and the lines where we live were so long that my wife read *Exodus* sitting in a service station line—not merely mean mild inconveniences of that sort, but will represent an attack on our underlying economic system. We all ought to be concerned about jobs, and indeed I think we are: for that reason, among others, I do not want to be put in the position of defending our "proceeding so quickly." Instead, I would like someone to tell me why the government has delayed our proceeding for the last two or three years, which is time we could ill afford to be delayed.

MAYOR BRADLEY: Mr. Haynes, perhaps we ought to ask Assistant Secretary Hughes about the delay, because those who serve at the state and city level, and those who are described as environmentalists, really had this debate on offshore drilling sharpened in the course of the last year, when the program and the environmental impact statement were thrust upon them in the latter part of 1974 with only a few weeks given them to respond to it. Perhaps if that environmental impact statement had been prepared years ago, and the states, cities, and environmentalists had been given the chance to evaluate it carefully, we would now be at a point where we could say this is a clear and justified decision. Even now the cities and states are being told this is a clear and justified decision, though the department acknowledges that it will have to make some changes in the initial statement. Apparently the department wants to proceed with the site-specific environmental impact statements even before there is a change in its earlier impact statement.

MR. HUGHES: That is correct. Now, I think I might first address the question, Why the delay? I think all of us through the 1950s and 1960s went merrily on our way, trying to avoid hard decisions on energy, assuming there would never be a problem with the U.S. energy supply. Perhaps, in hindsight, that was a bad way to go, but it tended to be the way in which government went at all levels. What sharpened the issue in everybody's mind was the change in the price of foreign oil from $3.50 to $11.00 or $12.00 a barrel, a change which caused energy costs throughout the country to skyrocket and produced massive disruption. The government responded to the disruption and the skyrocketing prices with an attempt to accelerate the program through an orderly legal process. Admittedly, we do wish to accelerate the program.

As for the relation of the environmental impact statement on the government's overall accelerated program to the various site-specific statements, and particularly the one for southern California, we believe that it is in the national interest for us to complete the final statements at all levels as rapidly as possible. There is no attempt to do any business in a closet. We recognize that if the final decision on the California lease sale is to be reached, we will have to work up to that decision with a particular process oriented only towards southern California—which is why we have issued a site-specific statement for southern California, and why we intend to hold public hearings in southern California in the near term. We do not think we should be held back from these hearings because we are revising our overall statement.

The decision that is most critical is the decision to hold the sale, not the decision to issue a call for nominations, and not the decision to issue an environmental impact statement in a site-specific sense. I recognize we have an honest disagreement on the question whether state and local governments have had ample opportunity to examine the program. In many cases that issue comes down to a traditional states' rights or local rights concern.

GOVERNOR BYRNE: Is Mr. Haynes suggesting that, for instance, we should have explored and developed the Baltimore Canyon trough off New Jersey years ago?

MR. HAYNES: Governor, to be perfectly honest, I did not learn of the exploration potential of the Baltimore Canyon until just a few years ago. Certainly I know our geologists had been looking up and down the East Coast, and I am sure all of our competitors' geologists had been looking up and down the East Coast, just as they have been looking all over the world; but the Baltimore Canyon became a "household word" in the oil industry only a short time back.

GOVERNOR BYRNE: I am not criticizing anyone and I know that, as a matter of fact, it would have been economically unfeasible to develop the Baltimore Canyon trough as recently as a few years ago.

MR. HAYNES: It probably would have. And I would like to make it clear that when we say "to develop the Baltimore Canyon," we realize that we do not know whether there is a teacup full of oil in the Baltimore Canyon. There may not be enough oil to fill a glass. What we are talking about is the potential for oil, and unless we get on with this program, we will not know whether there is oil out there at all.

GOVERNOR BYRNE: I would like to make one thing clear. I believe there has been an organized campaign by the oil companies to talk unrealistically about jobs and to make it appear that anyone who wants the orderly development of the Baltimore Canyon trough or any other offshore area is opposed to the preserving of jobs in the United States. I think that appearance is absolutely misleading. If Mr. Haynes would answer a few questions— such as when we should have explored and developed the Baltimore Canyon and what the oil industry has against the orderly development of the Baltimore Canyon or any other area with the potential of offshore oil and gas resources—we will be able to put the whole issue in perspective.

MR. HAYNES: May I answer those questions? First of all, Governor Byrne says that the idea of relating energy to jobs is part of an organized campaign, but I think this is the first time I have ever spoken on the subject publicly. I can assure you my remarks are not part of any campaign as far as I am concerned: I happen to believe very strongly that jobs and energy are related. Secondly, we have been talking about developing the Baltimore Canyon. Now my remarks a moment ago had to do with the general subject of offshore exploration. We are talking not only about the potential for exploration off the New Jersey coast, but also about potential for exploration all up and down the East Coast. We are talking about the Gulf of Alaska. We are talking about southern California.

We are talking about every bit of the United States's outer continental shelf where there appears to be any potential for oil and gas. We are not directing our attention only to the Baltimore Canyon, and I am not suggesting that anybody is to be criticized for not having gotten on with the Baltimore Canyon. All I am saying is that various groups have emitted a certain amount of criticism against the federal government for proceeding too rapidly, but that, in fact, the government has delayed too long. I can cite what the oil industry kept telling the public and the government about natural gas back in the 1950s. I do not wish to deal in hindsight, but I might like to say we told you so—we were talking about these problems as early as 1952 and 1953, and the problems we saw then have borne fruit. There is a shortage of natural gas in this country. It is serious, and it affects Governor Byrne's state among others.

GOVERNOR BYRNE: We do not mind criticizing the federal government. We do it all the time. But as it happens we are now criticizing the federal government's policy on the shut-in gas and oil wells that some of the oil companies may have.

MR. HAYNES: I am not criticizing the federal government alone. I am very much aware that one thing that has impeded accelerated offshore leasing has been industry accidents. Those accidents were very much to be regretted, and I think the industry has regretted them more than anyone else.

GOVERNOR BYRNE: You have improved your technology since Santa Barbara, I trust.

MR. HAYNES: There is no question that the oil industry has improved its technology, and the government has improved its operating requirements.

GOVERNOR BYRNE: All of this has been for the good.

MR. HAYNES: Yes, this has been for the good. But I

would like to get back to the question of these accidents
for just a moment. I think four accidents on 18,000 wells
is a pretty good record, though of course it is not perfect,
and we cannot guarantee a perfect performance in the
years to come. But I would say to Governor Byrne that
we all took a risk even in getting out of bed this morning:
we cannot live without risk. I submit that the risk of offshore drilling is an acceptable risk. If I am criticizing
the government, it is because I wish the federal government and the states and the cities would get together and
get on with the solution to this problem. We have been
talking about an energy program for months and months
and months, and we do not have an energy program. In
the meantime, this country has an energy problem, and
the oil industry would be irresponsible if it did not treat
these matters as vital.

GOVERNOR BYRNE: Mr. Hughes will tell you that we
have been getting together.

MR. HAYNES: If we can be of any help, we would be
delighted.

MAYOR BRADLEY: May I ask for a comment by Captain
Cousteau on these accidents, as you call them—on the
harm to the sea and onshore. I think you are as familiar
with these areas as anyone. What is your judgment?

CAPTAIN COUSTEAU: It is obvious that an oil spill is
a disaster. Nevertheless, the record of the industry shows
that a lot more damage has been done by the transportation of oil than by drilling. Recently, to take just two examples, spills from two supertankers, one around the
Cape of Good Hope and one in the Chilean Channel on
the southern tip of South America, spread devastation
over a big area. Transportation is responsible for a lot
more trouble than offshore oil drilling: there is no question about that. Nevertheless, an oil spill in a well that
is not far away from shore, though a local and temporary
disaster, is a disaster nonetheless. My principal worry is
that for the oil industry, as well as for nuclear plants, those

very ones who may endanger public safety or the environment are the only ones who have the right to control and assess what they are doing and how they are doing it. The oil industry is controlling its own safety—the atomic energy industry is controlling its own safety—and apparently no one else has the right to investigate. I would strongly recommend that if offshore drilling is to take place an independent agency be commissioned to control the safety measures.

MR. HUGHES: Mayor Bradley, I think it might be appropriate for me to respond at this point. I would certainly take exception to Captain Cousteau's remarks since the federal government does, in fact, control the industry.

CAPTAIN COUSTEAU: The Department of the Interior is not an independent agency.

MR. HUGHES: It is independent of the oil industry.

CAPTAIN COUSTEAU: Not so much. The department and the industry have the same interests. I believe some more independent group, such as the National Academy of Sciences, should commission a neutral organization to control the oil industry's safety measures in offshore drilling.

MR. HUGHES: Since the Santa Barbara spill, the federal government has made a number of changes in the operation of the offshore program. The number of inspectors has increased almost tenfold. The number of operating orders and safety requirements to which the companies are required to adhere has been considerably increased and the cost of safety to the industry has certainly escalated. Additional research has been accomplished in the area of spill containment and in many of the other areas associated with potential spills.

We recognize that there may always be an accident, so long as humans are involved in the oil industry or any other industry, but we think that probability is small at this point. We think that of the two alternatives, massive tanker traffic or increased offshore operation, offshore operation certainly has less potential for damage.

MAYOR BRADLEY: Mr. Assistant Secretary, do you believe that there is a minimal possibility of damage from oil drilling? Is the Department of the Interior willing to set up some kind of a liability procedure to protect those who may be damaged if, in fact, a spill occurs?

MR. HUGHES: Yes. As a matter of fact, within the administration we are now trying to set up the final wording of an unlimited liability bill which would establish a front-end fund so that if a community were subject to an oil spill—and we certainly hope this never comes about—there would be money available immediately to meet the disaster, to help that community pull itself back together. We think the climate in the Congress is certainly going to be very good for an unlimited liability bill. As I am sure all of us are aware, Congress is debating several other issues related to the outer continental shelf program. But this is one that we in the Department of the Interior strongly support. There will be a debate over the actual numbers involved, but it certainly makes good sense from our standpoint to establish the fund.

MAYOR BRADLEY: Do you think it is appropriate that the program be continued on its schedule before these decisions are made?

MR. HUGHES: We think it is appropriate for the program to be continued. In the next four to eight weeks, the Congress will probably come to some decision on general changes in the OCS program. We hope the Congress will take up the issue of an unlimited liability bill this term. Because of the time frames involved, even if we should have a sale in southern California in the early fall of 1975, it would probably be at least another year to three years before anything occurred in the way of drilling, so we believe there is ample time for such a provision to be enacted. If we waited until each of the concerns in the immediate OCS program were resolved and until we had buttoned down our national energy policy, it might be several years, which we believe would not be in the national interest.

CAPTAIN COUSTEAU: I have a major objection to this. Unlimited liability compensation would encourage irresponsibility. The safety measures would be much stronger if they were taken in advance of a spill. We always see the same kind of attitude: we prefer to clean the water when it is dirty than to avoid dirtying the water; we want to repair the damage done on the coast after it is done, instead of avoiding doing the damage. It is always the same.

Now there was something concerning liability that shocked me during the discussions yesterday. The question was raised, At what are you going to value the lives of birds? We discussed at length whether it was $1.00 or $10.00. That is nonsense. The life of a bird is invaluable—it is something that is precious in itself and that cannot be tagged with a price in dollars.

MR. HUGHES: I think it might be appropriate to ask Mr. Haynes to give us some idea what changes have been made in the way of oil industry safety since Santa Barbara. I know considerable changes have been made: the industry has not stood still technologically, and government awareness on this issue has certainly increased.

MR. HAYNES: I do not think time would permit me to enumerate all the things that have happened. I believe we should give the government a great deal of credit in this area. A little earlier Mr. Hughes mentioned the additional number of inspectors required and so forth, and I started to interrupt him and say that the number of reports we have to file has multiplied about a millionfold. Though I believe that is all to the good in its final effect. I am delighted to have a chance to say something pleasant about the government because it may be thought that I have been criticizing the government too much.

The oil industry has improved our technology as Mr. Hughes notes. We have spent millions—

CAPTAIN COUSTEAU: You see? . . . Hand in hand.

MR. HAYNES: —I would hate to think we were not operating in concert, Captain Cousteau.

CAPTAIN COUSTEAU: I would like you to operate in concert with the public, too.

MR. HAYNES: We are certainly trying to do that. We are working with the government and I have always thought that the government was representative of the public.

CAPTAIN COUSTEAU: Through a complicated system.

MR. HAYNES: Be that as it may, not to belabor this point, we have improved our training, we have spent millions in research and development, we have formed cooperatives to handle emergencies when they exist—thank goodness the cooperative in San Francisco has been called on only once or twice, most recently by the Coast Guard. We have made great strides, but I do not want to take the time to elaborate on all of them. Obviously we are spending a lot of money in research and development for our drilling techniques and platform designs. We are working on underwater completion. We are doing a lot in this area.

CAPTAIN COUSTEAU: Would you be against the National Academy of Sciences' looking into it?

MR. HAYNES: Captain Cousteau, I would not be against that.

CAPTAIN COUSTEAU: No?

MR. HAYNES: As a matter of fact, I think that when the government lays down regulations, operating restrictions, and so on, they would seek counsel and advice.

CAPTAIN COUSTEAU: Yes, and why not from the National Academy of Sciences?

MR. HAYNES: I think that is a fine idea. We would not
have any objection to that.

MR. HUGHES: We certainly have invited outside sources
to provide input into the program.

CAPTAIN COUSTEAU: And why not the National Academy of Sciences, which is the highest authority in this
field?

MR. HUGHES: What would be their exact function?

CAPTAIN COUSTEAU: To name the experts, to control
the reports, and to give their advice to the government.

MR. HUGHES: The academy does of course give advice
to the government. Is your proposal for the academy to
review the program?

CAPTAIN COUSTEAU: No. To review safety—only
safety. That is all.

MR. HUGHES: I think I would be opposed to asking the
academy to undertake an operational program of safety
and inspection, in part because it would be subject to
the same sort of criticism to which you are subjecting the
federal government.

CAPTAIN COUSTEAU: Why?

MR. HUGHES: Why not? We have scientists who work
in our department, too. Do you mean to imply that the
academy would have scientists who are a little more
objective?

CAPTAIN COUSTEAU: A little more independent, yes.

MR. HUGHES: But where does the National Academy of
Sciences obtain its funding?

CAPTAIN COUSTEAU: From various sources—not only from the government.

GOVERNOR BYRNE: May I ask a question? The suggestion has been made that we have an either-or situation—either supertankers or offshore drilling. Mr. Haynes, is that so? Would offshore resources eliminate the need for the supertanker?

MR. HAYNES: I think the easiest way to answer that question is to say that if we do not get on with offshore exploration and find some oil, and (I might add) produce it, then we are going to be more dependent upon foreign sources. Now, there is no other way to get foreign oil into this country than by tankers. There is at present no way to bring a supertanker—VLCC (Very Large Crude Carrier)—the so-called 200,000-ton class and above—into the United States, because we do not have a port facility to accommodate one. I think it is horrendous that we do not, because it means that the consuming public is paying a premium for transporting oil in smaller tankers. To answer the governor's question, we can say that if we import more oil, the only way to get it here is in tankers, and a lack of offshore drilling will increase the tanker traffic. I do not think our debate should be on the advantages of VLCCs as against small tankers.

GOVERNOR BYRNE: I did not want to make that the issue. I just wanted to make sure whether you think we are in an either-or situation.

MR. HAYNES: I do not think there is any question that we are going to have additional tanker traffic to this country.

GOVERNOR BYRNE: In any event?

MR. HAYNES: It is the only way to get foreign oil here.

GOVERNOR BYRNE: In any event?

MR. HAYNES: No. Obviously, to the extent that we can

find and develop oil offshore, that oil will displace imports.

CAPTAIN COUSTEAU: For a few years?

MR. HAYNES: Well, until we can get onto some of these alternate sources of energy.

CAPTAIN COUSTEAU: But we have not tried. We have not tried.

MR. HAYNES: Pardon?

CAPTAIN COUSTEAU: That is my entire point today. We would very much support the temporary bridge of off-shore oil drilling if it were a bridge to a monumental program for alternate nonpollutant sources, but such a program does not exist.

MR. HUGHES: Captain Cousteau, I think the government does have programs for alternate sources.

CAPTAIN COUSTEAU: Yes, but I happen to know that the programs are ridiculously small when compared to the amount of investment that is necessary. I have all the information here. We need a total of about $1 trillion in fifteen to twenty years. This, of course, is not to be paid for by the federal government alone; it has to be paid for by the economy as a whole. If such a program is not adopted, what then? The program of offshore drilling is just a temporary oxygen tent before we suffocate from lack of energy at the mercy of the foreign producers. That is exactly what is going to happen in twenty years.

MR. HAYNES: Captain Cousteau, the oil industry certainly supports a massive research and development program.

CAPTAIN COUSTEAU: I am not concerned with the oil industry's program.

MR. HAYNES: I know that. I just want to have it on

record here that we support a massive research and development program in this area.

CAPTAIN COUSTEAU: Yes, I know.

MR. HAYNES: So far as my company is concerned—I cannot speak for the whole industry, but I am confident that others are doing the same thing—we are spending sizeable sums on research and development in coal gasification, coal liquefaction, shale oil development, solar energy, geothermal energy, and things of that nature. It will still be ten or fifteen years before those sources become available to us.

CAPTAIN COUSTEAU: Fifteen to twenty years.

MR. HAYNES: I might add that the cost of these alternate sources of energy are nowhere near the figure that is being bandied about in the press and in Washington. But I can assure you that at $6.00 or $7.00 a barrel—and I am talking about 1975 dollars—there is not going to be a lot of shale oil development or coal liquefaction or coal gasification or tar sand development. It just would not be economically feasible, because the costs would be in excess of $14.00 to $15.00 a barrel.

CAPTAIN COUSTEAU: I am sorry.

MR. HAYNES: This is not to say we should not go into these things.

CAPTAIN COUSTEAU: No.

MR. HAYNES: But I think we should recognize that it is not going to be cheap.

CAPTAIN COUSTEAU: And these alternate sources of energy that are the only major ones now really being considered in the national program—such as coal gasification and to a lesser degree tar sands and shale oil— are either potentially heavily polluting sources of energy or land-destroying sources of energy. The only sources of energy that are absolutely pollution-free and are in a tremendous abundance in this country—geothermal and solar—are considered as utopian.

MR. HAYNES: I hope not, because we are doing some work in those areas right now.

CAPTAIN COUSTEAU: You largely said that yourself—that if you do not have the appropriate support from the authorities, they are utopian. The federal government must turn its program to nonpolluting renewable sources of energy. If it did so, I do not think there is one environmentalist who would not support a program of offshore drilling.

MAYOR BRADLEY: Mr. Hughes has said that the government does have a program and is proposing to spend money or is spending money in this kind of research.

MR. HUGHES: Right.

MAYOR BRADLEY: Perhaps he can quickly tell us how much.

MR. HUGHES: I think Captain Cousteau disputes the figure, but it is my understanding that the 1976 federal budget contains $53 million for solar energy research. We are also spending—

CAPTAIN COUSTEAU: That includes all forms of solar energy?

MR. HUGHES: That is correct. In addition, we are spending tens of millions of dollars in other fossil fuel research on the problems of coal gasification and liquefaction. We are certainly continuing to explore and experiment with geothermal resources. It is the opinion of our experts at this time that geothermal energy will never supply much more than a very small amount of the energy requirements of this country. But we are not giving up on it as an alternate source of energy.

MAYOR BRADLEY: One other question. Since there seems to be some suspicion of any other research group, whether it is government or industry, what would be the position of the Department of the Interior on the National Academy of Sciences' conducting the review of the program and making a report?

MR. HUGHES: We still have that question under advisement. We are not willing to make a commitment at this point to have such a study conducted if we have to suspend all activities in the program for some period of time, something that is normally implicit in the suggestion that the National Academy of Sciences carry out the review. We are certainly continuing to work with organizations like the National Governors Conference on the idea of a third-party shoreside impact study that would not be suspect because the Department of the Interior or indeed any federal agency had done it. Our basic intention is to conduct the orderly program laid out in our tentative schedule and get these answers along the way, rather than stopping everything now and waiting one to five years before resuming a program that requires a five-year lead time.

MAYOR BRADLEY: Thank you very much, gentlemen, for the illuminating discussion. And now we turn to questions from the audience. I should mention that the audience here includes a number of recognized experts in the field.

PROFESSOR WALTER MEAD, University of California, Santa Barbara: I have a question for Mr. Haynes. He spoke at length about the employment effects of another embargo. His point was that a second embargo would be a disaster. Then he said that the only solution—I think that was his exact wording—was offshore development. I think he might perhaps agree with me that offshore development is part of the solution, but not the only solution. I was a little concerned that there has been no discussion here about the possibility of crude oil storage inasmuch as a system of storage might be set up to provide us with one year of security or even two years. Could Mr. Haynes comment?

MR. HAYNES: When I made the comment that offshore drilling was the only solution, I think I said that it was the only medium-term or short-term solution for developing additional energy resources in this country. And I do stand by what I said. Now I did comment on the effect of an embargo. I think we may have been remiss in not addressing ourselves to strategic storage, which I think is an excellent idea. It is one, as you know, that the National Petroleum Council recommended less than a year ago, and it is one that my company wholeheartedly supports; it is part of the President's energy program, and I think it makes a lot of sense. I hope that the energy debate between the administration and the Congress will soon be resolved, and that when it is we will have strategic storage. I think strategic storage would certainly mitigate the effect of another embargo.

DON TILLINGHAST, Office of the Attorney General, State of Alaska: I have three quick questions for Mr. Hughes. First, I wish he would expand on the problems with the tanker/platform trade-off theory as it pertains to OCS operations off Alaska. Second, in regard to his discussion with Captain Cousteau on the objectivity or the lack of objectivity of regulation by the U.S. Geological Survey (USGS), I wonder why the Department of the Interior has refused to require application of the best commercially feasible technology for exploration and development, despite the recommendations of scientists, despite the recommendations of the Environmental Protection Agency (EPA), and despite recommendations of the Council on Environmental Quality (CEQ)? Third, if the Department of the Interior is as objective as the assistant secretary says it is, then why in the Gulf of Alaska is it continuing the practice of developing operating orders in consultation with industry, even though Dr. McKelvey, the director of the USGS, promised the House Government Operations Committee it would discontinue that practice and would give industry only the same role in the development of those orders as the affected states and the interested public?

MR. HUGHES: I will try to answer the three issues you raise. First, on the trade-off between platforms and tankers as it affects Alaska: I think we all understand that Alaska is a unique case. Alaska is certainly more than self-sufficient in fossil fuel resources; the state will be a net exporter to the other forty-nine states. Alaska has areas of high resource potential and several that we are interested in exploring and perhaps developing in the near term. Once those resources are located, platforms would in fact be constructed and tanker traffic would proceed from Alaska to the lower forty-eight to deliver that oil. If that was the point of the question about the trade-off, I can say that we know Alaska is a giver and not a taker of energy resources.

On the matter of the government's failure to recognize the best commercial technology and failure to require its use, I must confess that I am not acquainted specifically with the issue. I think we require a certain standard of excellence so far as the technology is concerned, but whether the cost-benefit trade-off requires the best technology is something we can explore and then inform Mr. Tillinghast or Governor Hammond.

As to our continuing to work with industry on putting operating orders together, that is something, again, that I am unaware has been a subject of great controversy. We certainly need industry input on the technology that industry is planning on using in a given area, and Alaska has a hostile climate, a climate in which new operating orders or variations of existing operating orders are going to have to be put together to ensure responsible resource development. But, again, we can be in touch with Mr.

Tillinghast or the governor to confirm or deny Dr. Mc-
Kelvey's commitment to a congressional committee.

LEONARD MEEKER, Center for Law and Social Policy:
I would like to address a question related to Captain
Cousteau's proposal both to Captain Cousteau and to Mr.
Hughes. I would like to ask Captain Cousteau, first of all,
what level of funding and effort he would consider appropriate today for pursuing the development of renewable
and nonpolluting energy resources, particularly solar
energy. I would like to ask Mr. Hughes whether the government takes a different view of the appropriate level
and effort today and, if so, what the reasons are for its
taking a different view.

CAPTAIN COUSTEAU: The answer to your question is
slightly complicated, but it is substantiated by research
done by several organizations in this country over several
years and by Eurocean, which I chair. We have consulted
with such well-known study groups as Battelle in Switzer-
land. Of course, the figures that I am going to give you
are within 30 percent, because it is very difficult to esti-
mate such advanced projects. In discussions of solar
energy up to now, the various governments of the world—
and I am not pinning this down to one government, since
they all do the same thing—have diverted the question
just to the heating and cooling of houses. The same thing
happens when environmental problems are raised: the
people divert attention from the real problems by inducing
their children to pick up beer cans. The real problem is
not only to make energy to heat and cool individual houses,
but to have huge plants making thousands of megawatts,
and for this we need huge sums of money.

Let me give a few figures. In order to make things
easy to understand, let us talk in terms of megawatts
rather than BTUs or watts. A large nuclear plant will
produce 1,000 megawatts. At the moment, the world's
consumption of energy is equivalent to the production of
17,000 nuclear plants. At an efficiency of one-ten-thou-
sandth of one percent, we could extract from solar energy

the equivalent of 76,000 nuclear plants. This country is exceptionally well located for solar energy, having both the magnificently solarized desert of Arizona and the Florida current, each of which, in different ways, indirectly concentrates solar energy in small places. One square yard of the Gulf Stream in the Florida Straits has twenty times more energy concentrated on it than one square yard in Arizona. We have there a preconcentrated —one might say prefabricated—source of energy.

What I would like to see is a commitment to build an OTE test plant of, let us say, 1,000 megawatt capacity. It would probably cost around $3 billion. And of course OTE would not be the only area of research. I think that the investment in the first year could be either by direct funding or tax rebates or city money with industry joining in: the joining in would probably be necessary because the first year's budget for geothermal and solar energy research and development together should be at least $10 billion. If developed properly, this program would probably cost almost a trillion dollars in fifteen years. This is the cost we have to face: this is the price of total independence. The United States of America could be a major exporter of energy—indeed there is no other country that could produce as much solar energy.

MR. HUGHES: I think I would have to begin by saying that it would be difficult to dispute the findings that some eminent scientists have come up with in putting that study together, but that any decision to spend massive amounts of federal money on a program like this (or any other program) must always be viewed through a cost-benefit comparison. By the way, I would like to correct a figure I used earlier: in this year's budget it is $153 million, not $53 million, for all sorts of solar energy research. Certainly this is not spending at the level that Captain Cousteau would desire; but it is the opinion of experts in the government, and, I am sure, of some of the experts in the National Academy of Sciences, that that is the best program that the government can operate for the near term. Nevertheless, we are certainly willing to look at

Captain Cousteau's proposal, even though the idea of adding $10 billion for solar energy research to the country's existing budget burdens, when the solar energy might not be available for many years, does not seem particularly attractive. We continue to believe that we are going to be dependent upon fossil fuels in the short term, and part of our belief stems from the calculation that the government should not spend nearly a trillion dollars for solar energy right now, though that is certainly a decision open to review every year.

CAPTAIN COUSTEAU: You just implied that I said something I never said. I never said that development of solar energy would preclude development of intermediate sources of energy. I am sorry that we are always talking about the short term, the short term, the short term. I must say, I think a large segment of the population is looking a little further.

PROFESSOR WILLIAM MENARD, Scripps Institution of Oceanography: I have a question for Assistant Secretary Hughes, who, I am afraid, is getting most of the questions here. There are several bills before the Congress proposing that the Department of the Interior undertake a program of exploration of the continental shelf—including drilling—in order to get information for policy decisions. I wonder if we could learn more about the department's position on these bills. I believe I understood him to say that the department believes there would be a delay in drilling, or in development of the shelf, if the department were to do the exploration. Could he explain why this delay would arise and how long it might be?

MR. HUGHES: Allow me to review the entire subject briefly. We have several objections to the idea that the government should directly or indirectly get into exploration.

First of all, government control of exploration implies that the government will decide where the holes are to be drilled in the search for oil. We have a major problem with that, because it would centralize decision making in the oil business either in one person or in one government agency, whereas under the present system it is decentralized among fifty or sixty companies that participate in the process.

All of us must keep in mind that the oil business is a speculative business, that the oil companies are in fact gambling each time they read their seismic information and make a decision to spend money on it. The most optimistic company in a given case will win the competitive auction and will have an opportunity to see whether its decision was correct. To reduce the number of those who analyze the information, the number of gamblers, if you will, to one group, the government, would not be in the national interest. It cannot help reducing the probability of finding oil. That is our first objection to it.

Secondly, it would require that the government absorb the front-end money required to drill holes. Drilling on the outer continental shelf will range somewhere between $1 million and $10 million per hole, depending on the type of climate where the drilling takes place. In the Gulf of Mexico the cost would probably be on the low end of the spectrum, while as we move up into the more hostile areas, such as offshore Alaska, it might be closer to $10 million per hole. The government would have to pay those front-end costs, and as the person responsible for the budget for my department, I cannot say I care much for the idea of going back to the Congress and saying, We just spent $50 million of the taxpayers' money and have fifty dry holes or twenty-five dry holes, and now I want another $50 million to drill more holes. I have the feeling there would be interminable delays in getting that additional $50 million.

Our system of government, especially when it comes to the relationship between the Congress and the executive branch, is not noted for its efficiency. As a matter of fact, the government was designed with built-in delays so there would be time for our different branches to reflect on decisions. We believe if we shifted to government exploration, there would be delays. Whether the delays would be good or bad is something we could debate but not here. The point is that we think any change that would make the government, either directly or indirectly, responsible for drilling would cause delay. We would have to decide who would make the decisions. We would probably have to create a new agency or to expand the U.S. Geological Survey to take on a new role. There would be a reluctance for whomever we made responsible to jump right in and make those decisions immediately. This would be a very proper reluctance I am sure, since there would be a substantial delay caused just by the shift-over.

All this is true, but our final and basic objection to having the government take over is that we think that private industry in this country is doing the job very well. We think we have the best oil industry in the world, and an industry adequate to handling the problem, and we see no reason to change the overall system. We have made several changes in the system; we are not saying that the program itself should be inviolate. We are proposing, for example, that there be a ban on joint bidding among major companies in order to get away from the charge that the major companies dominate the OCS program. We think the ban on joint bidding is a proper action for the government to take. I suppose that the large oil companies as a group object very strenuously to that sort of restriction, but we think it is in the public interest. But this is a change within the existing system, not a change of the system as a whole.

GOVERNOR BYRNE: Mr. Mayor, may I ask Mr. Hughes what Secretary Morton meant when on March 14 he testified in part as follows: "The National Governors Conference has called for a formal separation of the decisions to lease in the area and to allow actual development in that same area. I support the objective of the National Governors Conference proposal in principle and have asked my staff to determine the administrative steps necessary to put that policy into force without introducing undue delay in the development of the national energy resources. Our solicitor has informed me that such objectives could be accomplished."

MR. HUGHES: Secretary Morton did make that statement. We think that under existing law, the 1953 Outer Continental Shelf Lands Act, we have the legal ability to create a pause between exploration and development. We are now trying to establish what such a pause would mean to the interest on the part of the oil companies in investing massive amounts of money, time, and effort in the research they undertake before they offer bids on the OCS.

GOVERNOR BYRNE: Does this testimony commit the secretary to that concept?

MR. HUGHES: Yes, in a sense, it does commit the department to finding a way to create the pause. We do not subscribe to the idea that the government should conduct the explorations, and we think we can accomplish the same goal that you and other governors want to accomplish through existing law.

MAYOR BRADLEY: In further response to Dr. Menard's question, Mr. Haynes wants to make a comment from the industry's point of view.

MR. HAYNES: I am delighted to hear the assistant secretary observe that he believed the federal government should not become involved in oil exploration. I want to commend him for that decision, because I happen to agree with it wholeheartedly for the reasons he mentioned and perhaps one or two more. Perhaps I am a little bit old-fashioned, but my basic view of government is that government should do for the people what the people are not capable of doing for themselves—if it should be done at all. To me, that is fundamental—so fundamental that it may sound trite.

In all humility, I think the oil industry over the past several years has done a very commendable job in bringing an abundance of energy at a reasonable cost to the developed and developing countries of this world. I think that getting the federal government into oil exploration would be an absolute disaster. I think it would delay off-shore development for years and years.

Now, though I do not really believe that the advocates of the government's becoming involved entirely mean what they say, if we were to take some of their statements literally, we would find that they are suggesting the government explore for oil, make an absolute determination as to the amount of oil that is in a particular basin, and then open the basin up for bidding. In other words, the industry would in effect be buying fuel oil in the tank. I would like to remind anyone who is interested in this subject that the Gulf of Mexico is one of the most prolific oil-producing basins in the United States. We have been producing oil there for twenty-five to thirty years, and we are still exploring for oil in that particular area. If the statements are taken literally, these advocates of government involvement suggest the U.S. government spend twenty-five years determining how much oil is in an area before it decides to develop it. Of course, they cannot really mean what they say, but I think this is a point we must consider. Even for a given structure, a given salt dome or whatever geologic configuration there may be, we do not define in absolute terms the total reserves until we have drilled a few exploratory holes, a few delineation holes, until we have installed some platforms, until we have drilled development wells, and have obtained some production history. Even then we find oil years later in deeper horizons. Now, I am not trying to confuse this issue with details. I want to be sure that people understand that, if we were going to separate exploration from production or development, we would have to be very careful how we defined exploration and production.

Furthermore, if anyone really believes that the federal government could do a better job than private industry, I would like to commend to them a very quick look at who has found oil in the free world. Basically, it is the international U.S. oil industry that has found oil. There are a number of national oil companies all over the world, and with the possible exception of what the Mexican oil company found just a few months ago (after looking for some twenty-five years, I might add) and possible success by the Petrabra Company in Brazil very recently (I might add after a long effort), the world's national oil companies have not found enough oil to fill a glass. I think this reminder is needed so that we will keep the matter in the proper context.

I want to fall back to my original point. The oil industry has done a commendable job. We have made some mistakes. We regret the mistakes. We have a lot to be proud of, and I submit we can continue to do this job better than the government could do it. If anyone questions that, I would tell them to look at the postal system or Amtrak, to take just two examples.

GOVERNOR BYRNE: May I ask, if the industry is so anxious to keep the government out of this, why it is also so anxious to take tax write-offs on dry holes?

MR. HAYNES: My firm—like others—happens to be a business organization, and we have some 280,000 stockholders who have invested about $8 billion in our company.

GOVERNOR BYRNE: My only point is you are involving the government at least in your failures.

MR. HAYNES: We are involving the government the same way as the man down the street who has a factory and depreciates his plant and equipment. We are a profit-making organization: we have the incentive to make profits and if the government will give us the economic and political climate, we can get the job done.

RICHARD SNODY, National Management Association: I noticed in Governor Byrne's opening statement that he seems to be somewhat disenchanted with the fact that the offshore resources would be divided up and handed to the highest bidder. What alternative means of exploration can he offer?

GOVERNOR BYRNE: I do not know why Mr. Hughes or

Mr. Morton divided the offshore oil leases into such huge blocks. I do not understand where they got the number of blocks, how they got it, and how they justify it. Second, I am not sure yet how they are going to price those leases. I would be strongly in favor of a pricing system that did not "front load" those leases with huge bonus bids; I would much prefer a system that required a minimum commitment at the beginning and a higher commitment in the royalties during production. I think Senator Bentsen of Texas has a bill that would go a long way toward doing that, so that other companies and even the states might— on their own—join in bidding for those leases, do the exploration, and make the judgments as they went along. My main reservation is this huge financial commitment at the "front end" of the current leasing/exploration/ development system; there are many alternatives to that.

WILLIAM REILLY, The Conservation Foundation: I think a good deal of the concern about offshore oil involves the onshore impact. In the environmental impact statement issued last year on the offshore program, about 20 out of some 2,000 pages were devoted to onshore effects. But we have just completed a study of the onshore effects in Scotland. We discovered that they are considerable—tank storage farms, refinery construction, pipelines, production, rigs, and all the rest of the things and activities that go with them.

Now, earlier this year the Department of the Interior proposed legislation to site energy facilities, in some instances over and against the wishes of state and local governments. My question is, What is going on in the department that would reassure people like Governor Byrne and the mayors as to the thought that is being put into the serious onshore problems of offshore oil and gas development and would cause them to want to give the department this authority?

MR. HUGHES: We would certainly admit that the department's role in the onshore impact question has changed over the last year to eighteen months, as we have thought in more detail about the potential impact on states other than our traditional Gulf Coast areas, where the program has operated for twenty-five years. As I mentioned, within the department we are now engaged in a dialogue with the National Governors Conference staff on what third-party entity might be able to conduct an objective study on potential onshore effects. I think all of us should keep in mind the fact that the onshore impact would vary with the locality. For example, the tiny village of Yakutat on the Gulf of Alaska has 400 people and would look forward to perhaps several thousand people coming in over a short time, should a decision be made to have a sale, and large amounts of oil be found in the gulf. Needless to say, the state of Alaska would have a then tremendous problem in establishing basic infrastructure, roads, schools, sewage systems, and so on.

The impact in southern California would certainly differ from the impact in Yakutat, and the impact in New Jersey and other East Coast states would be still different. We are not trying to minimize the importance of the onshore impact from offshore development. In fact, we are trying to get a better data base upon which to encourage some changes in the system. I think the press has reported that we in the department have been reviewing revenue-sharing options, which might provide other sources of financial support to the states; but there are a lot of unknowns in the process. We hope that by working together with the states and perhaps with the National Academy of Sciences—which may be the body that we ask to do this study—we can get some outside expertise on a very complex problem. But we do not think the matter of shoreside impact is serious enough to lead us to stop the entire program for two or three years. We think there will be enough time in the course of the program to review the problem and see what we need to do to help mitigate potential adverse effects.

JOSEPH EDMONSTON, Sierra Club: The Standard Oil Company of California has been asking for a supertanker port near the Morro Bay estuary, and has also been supporting OCS drilling in southern California. This has led some people to question whether we really will have a trade-off between tanker imports and OCS development. I read several days ago an analysis suggesting that what the oil companies are trying to do is develop the OCS and tanker imports to the extent that they can then justify sending North Slope Alaska oil to Japan. I would like Mr. Haynes to comment on that, please.

MR. HAYNES: I would be delighted to comment on that. Let me take the last part of your question first, if I may. Our company has not been as successful in its exploration activity on the North Slope as our competitors have been in theirs, and I envy them. The fact is that we do not have enough production on the North Slope, or will not have enough production out of the Prudhoe Bay complex, to export anywhere except to our refineries in California. I think that the fact that we are talking about a superport up near Estero and about the OCS and that we do not have any Alaskan oil export probably answers your question. I think the industry has made it abundantly clear in the press and otherwise that it has no intention of exporting Alaskan oil to Japan or anywhere else except to the West Coast of the United States.

Now, not only have we talked about a superport in California, but we are also part of a project to build a superport off the coast of southern Louisiana. The industry is also involved in trying to put together a superport complex off Texas. There is also some discussion about offshore Mississippi. There is also some discussion about the East Coast, though I have forgotten the exact locations. All this superport construction is in the interest of trying to bring oil to the American consuming public at the least possible cost, because the economies provided by shipping oil in these large tankers is significant and would be reflected in the cost to the consuming public.

I do not think our position here is inconsistent at all with our desire to get on with exploration on the outer continental shelf, because if we were to develop the oil there, we would minimize our dependency on foreign sources of energy for the next decade or two, but we would still be importing. I think it behooves us to import in the most efficient manner.

BARBARA HELLER, Environmental Policy Center: I would like to address two questions to Mr. Hughes and Mr. Haynes and perhaps get Governor Byrne's reaction to their answers.

Mr. Haynes stated earlier that the industry has been asking for accelerated leasing for years. However, in the most recent lease sale in the Gulf of Mexico last month, out of all the tracts that were offered, the industry bid on less than 30 percent. There are well-documented and generally agreed-upon shortages of capital, labor, and drilling equipment in the industry. These shortages have been reported by the trade journals and *Offshore Magazine*, as well as by the Federal Energy Administration. The question is, How is the industry going to cope with accelerated leasing?

Mr. Hughes spoke about the orderly process which the department has been pursuing on OCS development. This process, however orderly it is now, has come about after very vigorous public opposition, and, I think he will admit, in reaction to opposition from the governors as well as from citizen groups. I would be interested in knowing how the department is now trying to involve the public and the states in the decision process. They have issued proposed operating orders for the Gulf of Alaska which were published in the *Federal Register* about a month ago. Yet the material on which operating orders are supposed to be based has not yet come out in the environmental impact statement for that area. It seems to me that is not an example of involving the public in planned development.

MR. HAYNES: I think the thrust of your first question was that on the one hand we are advocating offshore leasing and on the other hand industry only bid on 30 percent of the leases offered in offshore Texas a few weeks ago, and that appears to be inconsistent. I think the only way I can answer that point is to say that the oil industry, of course, is a risk-taking, but also a profit-making, enterprise. Certain areas are put up for bids. Various companies nominate certain tracts, and the government decides which tracts are put up for bids and which are not. We have a certain amount of seismic and other information which we have acquired through our own efforts. We sit down and make the best judgment we possibly can as to the potential of a particular block. It so happened that we only bid on a few blocks, because we thought those were the only blocks that had a potential that would justify our investment and our efforts. After all, justifying investment is the name of the game. I can only conclude that the rest of the industry looked at it exactly the same way, and that therefore a number of these blocks were not bid on. Incidentally, a lot of this area was one in which very little activity has taken place in the past.

I would like to remind you that offshore in the gulf in 1974 the oil industry bid about $2.48 billion on leases in excess of the second high bid. I make that point only to suggest that the bidding process is of some economic value to the federal government. As a matter of fact, since about 1958 the oil industry has spent $60 billion offshore. The government has derived $15 billion from the industry, and the industry has a gross income of just about $15 billion. We are a long way from breaking even. But I repeat that, if the government will put up the blocks that have exploration potential, the industry will assess the risk and the possible gain, and we will get in there and play the game, and find some oil, too.

MR. HUGHES: To answer the second part of your question, we do not think it is fair to characterize our program as being created in response to public pressures. An orderly program has been in existence for many years now. We admit we are making some modifications to the program, and we are making a deliberate attempt to work more closely with governors and mayors who are involved because of the concerns of their constituents about offshore development in areas where it has never occurred. We believe that the system has needed some modifications and we have adjusted our programs in an attempt to meet the concerns of the states and the people of the country. But the actual leasing process is pretty much the same as it has been. We believe that public involvement, either directly or through public hearings, is an important vehicle for the department as well as the rest of the federal government to gain input into the program, and we also think increased cooperation with the states is important.

Secretary Morton said in his meeting with Governor Byrne and his fellow East Coast governors in January 1975 that he would invite the governors to look at the final decision document on whether to have a sale. That would certainly be a step forward in the government's opening its process to the public. A governor is the public at large or the public as represented by state and local officials. On the matter of our OCS orders for Alaska being published before the environmental impact statement is written, I do not have a solid answer except that ours is a dynamic system. We certainly do not mean to imply that there will not be more OCS orders if something new and unusual is found through environmental impact analysis in Alaska. But we need to get things done in an orderly fashion. If we wait until everyone's objection to the program, no matter how small, is taken care of, then the process will be stalled for a long time, and we do not believe that would be in the public interest.

GOVERNOR BYRNE: We do not call it pressure: we call it discussion. I appreciate Secretary Morton's having had the discussions with us. And he did say in his testimony before the Congress on March 14 that it was as a result of our discussions that he invited the governors to participate in the evaluation of the environmental impact of offshore oil. I have often said that I appreciate the willingness of Secretary Morton and of Mr. Hughes to have these discussions with us, to learn our views firsthand, and to consider our suggestions.

ROBERT CAHN, The Conservation Foundation: My question is for both Assistant Secretary Hughes and Captain Cousteau. We heard in the discussion that we need offshore oil immediately. Captain Cousteau said that if we use offshore oil up now in a time of peace, we might not have it in a time when it was needed strategically. I want to ask the question, What is our responsibility to future generations for using up this immense source of oil in fifteen to twenty years, when we do not know whether there will be other sources of energy available that could be beneficial, or not harmful, to people? A related question is, What is the government's policy on intense conservation of energy, which may be the solution to the problem of our doubling the rate of use every few years and so using energy up rapidly?

CAPTAIN COUSTEAU: I must insist that I said that I was against using up this oil if the question of replacement were not studied at the same time, because otherwise the sacrifice of the offshore oil would be useless. But if the appropriate research and development were carried out, I would support offshore development, because I think we do need to use offshore oil, provided we have a replacement for the future. Let me make that clear. This is a constructive approach. It is not a negative approach.

MR. HUGHES: I think we recognize that a decision could be made to save our additional domestic resources in the short term for some future generation, but that seems to present us with the prospect of massive dislocation and disruption in our present economy. If, in fact, we decided that we want to use half the energy next year that we used this year, that could be accomplished: we can in fact do almost anything we want in this country, but there will be a cost. We think it is in the national interest to proceed in an orderly fashion to make domestic resources available over a period of time for the needs of the nation, hoping that our research programs over the long term on fossil fuel gasification and liquefaction and the more exotic forms of energy supply such as geothermal, solar, and nuclear will produce energy that will come on line to take up most of the burden as we reach the year 2000. But that decision could be changed. The Congress in its wisdom could make a decision that we would not proceed with offshore development for some period of time, although there would be a definite price for that decision. We believe that the program the federal government is pursuing now in cooperation with other levels of government is the right program, but of course any decision in our society is open to question.

On the matter of conservation, I would say that since last year we have taken about a million barrels a day out of the consumption patterns in this country, and that, unless we go to legislated items like those proposed in the President's energy package—such as tax rebates for insulation and other monetary incentives for conservation—we will not get much more in the way of voluntary conservation. If we get into mandatory conservation, saying for example that the people can no longer have electric houses but must use some other source of energy, we have the prospect of a massive dislocation of the economy. Certainly we are open to suggestions, and the administration is continuing to review with the Congress several variations on a national energy plan that would try to bring more conservation into the picture. But conservation alone cannot solve the problem. Those who would say that we can simply decide we are not going to have any increase in energy consumption have not been talking to the American public.

CAPTAIN COUSTEAU: I think a lot more could be saved through energy conservation. It was said yesterday that the increase in the price in energy will first hit the poor, which is a very solid statement. Why not establish a tax on a sliding scale? Each home would have a low tariff for the first so many kilowatts, and then a much higher rate for additional energy, for those who can afford it. This would induce the rich to shut down their electricity about which they do not care today. But if we increase the price of electricity for everybody, we are going to hit the poor people. A sliding scale is a must.

JAMES PARDAU, Assembly Resources and Land Use Committee, California Legislature: Back to the question of onshore facilities. The California Coast Commission is in the final stages of preparing its coastal zone plan for the 1976 legislature. Of course, it is going to be a tough land-use plan, and onshore facilities are probably going to be an element in it. I would like to ask Mr. Hughes whether onshore facilities are required for the kind of OCS drilling the federal government is considering in California, or are there sufficient facilities already in place in California? If the California Coast Commission should deny permits for additional major facilities, what would that mean? Would there be a strong push by the federal government to try to preempt the state in this area?

MR. HUGHES: There are no detailed answers to the question you pose. Certainly there may be a need for additional onshore facilities in California. Whether that need could be taken care of through the expansion of existing facilities or whether new ones would have to be created is not known, nor is it something we can predict today with any degree of certainty, when the actions we are talking about may create pressures five, six, or ten years from now. As to the possibility that the coastal zone plan might either directly or indirectly preclude energy facility siting, I can say that should this become a nationwide phenomenon, it might require that the federal government review whether the national interest is being served when a local area denies energy facility siting. An energy-siting bill has been discussed in the Congress at length over the last five or six years. Whether such a bill will pass during the current session is certainly open to speculation, but we are concerned about those who would say that the federal government ought to wait until everyone's coastal zone plans are completed. There is no obligation for a state to complete a coastal zone plan. States are encouraged to do so under coastal zone legislation, but it takes a long period of time to come up with a plan. We hope, of course, that any such plan will be in place by the time any onshore impact is a reality.

MAYOR BRADLEY: We are going to have to come to an end. We are grateful for the stimulating discussion and for the questions from the audience. We certainly want to thank our panel: Mr. Haynes, Captain Cousteau, Governor Byrne, Secretary Hughes, thank you very much on behalf of AEI's National Energy Project.